CREWEL

NEEDLEPOINT

CROSS-STITCH

WEAVING

KNITTING

D0642770

THE TEXTILE COLORIST

THE TEXTILE COLORIST

Faber Birren

with illustrations by Shirley Riffe

Presenting a harmonious series of color effects for needlecraft, embroidery, cross-stitch, crewel, knitting, crochet, macramé, weaving, fabric printing, batik, quilting, rug hooking, and other textile arts, arranged by one of America's leading color authorities

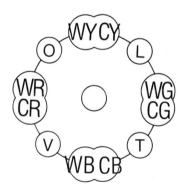

VNR VAN NOSTRAND REINHOLD COMPANY

NEW YORK CINCINNATI TORONTO LONDON MELBOURNE

Copyright © 1980 by Van Nostrand Reinhold Company
Library of Congress Catalog Card Number 79-27478
ISBN 0-442-23854-1

All rights reserved. No part of this work covered by the copyright hereon may be reproduced or used in any form or by any means—graphic, electronic, or mechanical, including photocopying, recording, taping, or information storage and retrieval systems—without written permission of the publisher.

Printed in the United States of America

Published in 1980 by Van Nostrand Reinhold Company
135 West 50th Street, New York, N.Y. 10020, U.S.A.

Van Nostrand Reinhold Limited
1410 Birchmount Road, Scarborough, Ontario M1P 2E7, Canada

Van Nostrand Reinhold Australia Pty. Limited
17 Queen Street, Mitcham, Victoria 3132, Australia

Van Nostrand Reinhold Company Limited
Molly Millars Lane, Wokingham, Berkshire, England

16 15 14 13 12 11 10 9 8 7 6 5 4 3 2

Library of Congress Cataloging in Publication Data

Birren, Faber, 1900–
    The textile colorist.

    Bibliography: p.
    Includes index.
    1. Color in textile crafts. 2. Color in art.
I. Title.
TT751.B53   701'.8   79-27478
ISBN 0-442-23854-1

CONTENTS

On Entering the World of Color

The great French painter, Eugène Delacroix, a master of color, once wrote: "The elements of color theory have been neither analyzed nor taught in our schools of art, because in France it is considered superfluous to study the laws of color, according to the saying, 'Draftsmen may be made, but colorists are born.' Secrets of color theory? Why call those principles secrets which all artists must know and all should have been taught?"

I concur with this opinion. Although the art of color harmony does not need to be strict or academic, there are certain basic truths and observations that should be heeded by anyone who seeks facility with color. In music, scales, keys. and counterpoint are fundamental to composers as well as musicians. Their laws should be known, even to the artist who plans to break them. People go to school not merely to learn but to rise above learning, to apply the wisdom and knowledge of others to original and creative ends of their own.

Many persons have a "good sense" of color. This is particularly true of women. They are probably more sensitive to color for the simple reason that, while colorblindness among men has a frequency of about eight percent, among women such deficiency is less than one-half of one percent. Women have also traditionally been involved with color. Handweaving, for example, has been women's work since history began. And it still is.

This book is at once for beginners, for advanced students of color, and for the most sophisticated! This writer is an old hand in the field, having published many books on color and having edited the works of such writers as Albert H. Munsell and Wilhelm Ostwald, M. E. Chevreul and Ogden N. Rood, all top authorities on color. During several decades I have had a conscientious respect for the good thinking of others, trying my best to draw from history, past and present, principles of color harmony and color beauty that to me constitute the foremost creative efforts. I have profited well from the formal views of profound color theorists and the more liberal and spontaneous expression of leading artists.

The conclusions I have reached concerning the secrets of color harmony are presented herewith in both illustrations and text. Harmonies are built around color circles, color triangles, and what have been considered "natural laws" of harmony. Much of this has been simplified to avoid the pedantic. The "laws" set forth, many of which are illustrated in the color plates, are all properly concordant—as the reader may judge.

Later in this book attention is devoted to the color expression of great artists, both as written and as exhibited in their paintings. They provide an education in color that is more related to emotional feeling than to rational deduction. Thus the reader will find that color harmony has been treated both as it may be *willed* by the formal theorist and as it may be psychologically *experienced* by the endowed artist.

There is one caution. As human beings, with complex psychic and highly personal responses, some of the color schemes included in the color plates may be liked by some people and disliked by others. People often respond to color in the same way as to their religions or political beliefs. It is common to find some souls who like red or yellow or blue and others who can't stand these very same hues. The moral here is that color harmony should be an intimate art and should reflect the spirit of the person who combines colors in arrangements that strike a particular fancy. One may pay attention to teachers for general instruction, but beyond this one should listen to voices that come from within and through them be uniquely and privately creative.

About Color Circles

It is perhaps fair to begin a discussion of color circles and charts with Sir Isaac Newton. Around 1666 he made the discovery that all colors were contained in natural light and could be separated by passing a beam of sunlight through a prism. Newton designed a color circle with seven hues: red, orange, yellow, green, blue, indigo, and violet. Seven was magic for Newton, and he allied his choices to the seven planets, the seven notes of the diatonic scale in music, and the rainbow. Red stood for C, orange for D, yellow for E, green for F, blue for G, indigo for A, and violet for B, while the seven colors of the rainbow became the delight of poets and have been romanticized ever since by them.

But six was better than seven (or other numbers of primary hues, as will be noted). Although the natural spectrum extends from red to violet, Newton brought the two ends together to form a circle and thus set a pattern that was to be followed.

Around 1730 Jacques-Christophe Le Blon made the ''amazing'' discovery that the primary colors were red, yellow, and blue and that mixtures of them in paints, dyes, and inks formed most other hues. This finding led to full-color process engraving and printing many years later. In this Le Blon was a great pioneer.

A hundred years after Newton, around 1766, Moses Harris, an English engraver and entomologist, presented to the world the first-known example of a color circle in *actual* color. The monograph containing it was dedicated to Sir Joshua Reynolds. Harris spoke of prismatic colors (pure) and compound colors (muted). His main circle of prismatic colors was composed of red, yellow, blue; orange, green, violet; and twelve intermediate hues. For his compound colors (known today as tertiaries) he mixed orange, green, and violet to form brown, olive, slate, and other intermediates, all having a subdued quality.

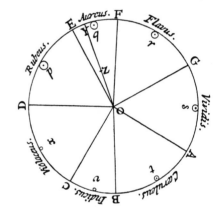

NEWTON

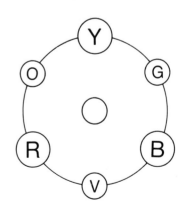

HARRIS

9

Although the Harris treatise was not and still is not readily available, it was known to British painters such as J. M. W. Turner, who drew modifications of Harris charts.

With Le Blon and Harris the tradition of the red-yellow-blue color circle had its beginning. It was later championed by such authorities as Goethe (1810), Chevreul (1839), and George Field (1817). For a while it became known as the Brewsterian Theory after Sir David Brewster (1831). In America it was the preferred circle of Milton Bradley (1890) and Louis Prang (1893), who manufactured art materials, Herbert E. Ives, and Arthur Pope (1928). At the Bauhaus it was sponsored by Kandinsky, Klee, and Itten. It is still basic to art and color education in the United States, and virtually every living painter is quite familiar with it. Right or wrong in concept, it has stood the test of time.

The reader is probably aware of the fact that color primaries can differ depending on the medium used. With light the primaries are red, green, and blue-violet. These are the colors of the phosphors on a TV screen. If red and green are emitted, yellow is seen. Red and blue-violet form a magenta red. Green and blue-violet form a cyan blue. Various other mixtures (called *additive*) form other colors, including flesh tones and white.

The additive colors in mixtures form the subtractive colors found in process printing. Magenta, yellow, and cyan (blue) are the usual transparent inks seen in magazine and book illustrations. These are *subtractive* and form other colors including black when intermixed. The color circle of the Munsell system to some extent respects both additive and subtractive primaries. The color circles of A. H. Church and R. A. Houstoun of England featured red, green, and blue-violet primaries, as did the circle of Wilhelm von Bezold (1876) and Michel Jacobs of the United States (1923).

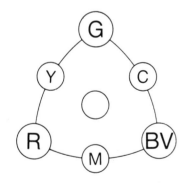

ADDITIVE

There is yet another set of primaries, which refers to vision and the optical mixture of colors. Four colors are involved—red, yellow, green, and blue—for they are primary sensations in human vision. In effect, these four colors are individual and do not resemble one another. However, all other colors bear a resemblance to them.

Curiously, in the fifteenth century Leonardo da Vinci wrote of the so-called visual and psychological primaries, and he allied them to the ancient concept of the elements: "The first of all simple colors is white, though philosophers will not acknowledge either white or black to be colors; because the first is the cause, or the receiver of colors, the other totally deprived of them. But as painters cannot do without either, we shall place them among the others; and according to this order of things, white will be the first, yellow the second, green the third, blue the fourth, red the fifth, and black the sixth. We shall set down white for the representative of light, without which no color can be seen; yellow for the earth; green for water; blue for air; red for fire; and black for total darkness."

Many years later (c. 1870) a German psychologist, Ewald Hering, assumed that the eye responded primarily to red, yellow, green, and blue, and he based a theory of color vision on them. Hering's conclusions were promptly accepted in the field of psychology and are still respected. Around 1916 Wilhelm Ostwald, also German, built a color circle around red, yellow, green, and blue primaries and conceived an entire system of color to include them. (His theories of color harmony will be discussed later.)

Finally, there is the color circle of Albert H. Munsell, which contains five key hues—red, yellow, green, blue, and purple. Five was an arbitrary choice, for he sought to formulate an entire system of color in decimal terms.

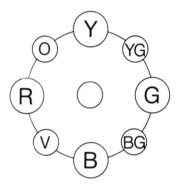

OSTWALD

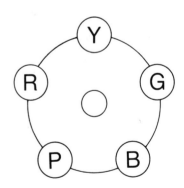

MUNSELL

Still another color circle is shown in Plate I, and justification for it is duly given. Color circles may be of numerous designs and patterns. The scientist may have certain technical factors in mind, and the artist may have other and more empirical views. As far as color harmony is concerned, all color circles have merit. However, because beauty is psychological and emotional in nature, any fast or restrictive rules may fail to apply. What may be attractive to one person may be rejected by another. Beauty indeed is in the eye of the beholder, and beholders are not all alike. As this book persists in emphasizing, color harmony should be a personal art. The purpose of this writer is to point out various paths and directions that he hopes will lead to delightful ends, all respecting the temperament and predilections of the reader.

The Color Triangle

Goethe arranged colors both in a circle and on a triangle (1810). It remained for Ewald Hering to reach the conclusion that a triangle was a natural form for color, with pure color on one angle, white on the second, and black on the third. With different triangles devoted to different key hues, all modifications (tints, shades, and tones) could be plotted and measured in the space within the triangles. The Ostwald system was based on this premise.

In 1937 this writer designed the simple triangle shown separately. Its primary forms are pure color (hue), white, and black. Its secondary forms are tint (pure color plus white), shade (pure color plus black), gray (white plus black), and tone (pure color plus white and black).

This color triangle plots the world of color as it is sensed and experienced in human vision. The interpretation is probably made in the brain. Plainly stated, every surface or object color seen in life may be classified in one of the seven forms of the triangle! That's all there is. There are seven ''boxes'' of perception; and, as different colors are seen, they are conveniently dropped into one of the boxes. Although the physicist may speak of millions of colors and color frequencies, the visual process strives for simplification. It struggles against complexity in favor of quick, general identity.

Perhaps the color triangle can be looked upon as a diagram that charts ''natural laws'' of harmony, for straight lines in any direction on it will lead to concord. This will be demonstrated by the color schemes shown in several of the color plates.

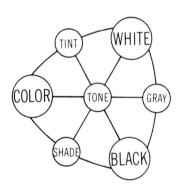

COLOR TRIANGLE

Plate I.
The Color Keyboard

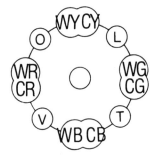

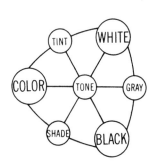

Look upon the color circle shown on page 17 as a keyboard on which basic chords may be struck and from which an infinite number of refined harmonies may be composed. This color circle is a special one. It has key elements in the four psychological primaries—red, yellow, green, and blue—plus their secondaries—orange, lime, turquoise, and violet. However, as will be noted, the primaries are divided into a warm red and a cool red, a warm yellow and a cool yellow, a warm green and a cool green, and a warm blue and a cool blue. This allows for greater expression with color, as effectively demonstrated in Plate II.

Many of the color schemes of this book play upon the chords of the color keyboard. There will be constant reference to it from plate to plate. The reader may likewise let fancy roam about the circle, follow principles described in this book, or plot combinations based on individual feeling.

The colors across the bottom of Plate I includes examples of each of the forms of the color triangle: tints, tones, and shades. They coordinate fairly well with the bright hues of the color keyboard. Any of the modified colors (tints, tones, and shades) may be combined with the bright hues or with each other to follow the natural paths of the triangle; such paths are duly described in this book. The color plates also present color schemes derived from the circle and the triangle and make use of all the colors shown in Plate I.

The yarns chosen to make up the plates have been purchased commercially, and it may be assumed that the reader will be able to match or approximate them. There are thirty colors, in addition to two grays, black, and white. As exceptions pale versions of pink, coral, cream, jade, and azure-blue tints have been used on some of the plates; where this has been done, the accompanying text will so explain.

Plate II.
Palettes of Distinction

Many color schemes devoted to embroidery, crewel, crocheting, knitting, weaving, and other textile crafts tend to have a similar appearance. Many exhibit commonplace arrangements of red, orange, yellow, green, blue, and violet, plus simple tints of these and hence look very much alike. Good color arrangements should have distinction and character. The overall impression should express a unique rather than a conventional personality.

The three color schemes (palettes) shown on page 18 bring together all the twelve bright hues of the color keyboard shown in Plate I. But the three combinations are quite different.

Scheme (A) combines warm red, warm yellow, warm green, warm blue, white and black. All bright (pure) hues automatically harmonize with white and black, as they include the primary forms of the color triangle.

The warm palette of (A) has a sunny quality, which reflects much of the art of the past. The cool palette of (B), on the other hand, brings together cool red, cool yellow, cool green, and cool blue, together with white and black. This has a fresher and crisper quality and a more modern and contemporary look. The exotic palette of (C) brings together orange, lime, turquoise, and violet and has a definite exotic and oriental look.

Study of human color preferences has revealed personality types that favor each of these three palettes. People with conservative taste tend to favor the warm palette. Those who are more extroverted, who enjoy newness and innovation, tend to favor the cool palette. Some who like the exotic palette are inclined to have mystical views, to be attracted to oriental philosophies rather than the conventionally familiar.

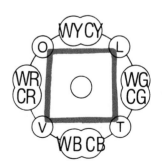

The reader may disagree, but the point can be made that color schemes, like foods, can be prepared in various ways to delight persons of differing gourmet fancies.

Plate III.
Tints, Tones, and Shades

Plates II and III make use of all the colors in the color keyboard shown in Plate I, including white, black and medium gray. Yet all the color schemes shown are distinct because they arrange colors in different ways and in different forms.

In scheme (A) of Plate III, shown on page 19, the six tints of Plate I are lightened with white, forming pink, coral, cream, jade, azure, and lavender. Tints have a natural affinity with white because they contain it. Colors of the same given form—tints, for example—are inherently harmonious. In the case of tints, this is due to their common white content. However, formal harmonies among tints are found in the upper schemes of Plate V and Plate VII.

Tints have a delicate, flowerlike charm and are springlike and young in spirit. (Pure colors are strong and more summerlike.) Tints are simple, unsophisticated, and chaste and have long been associated with much that is youthful in quality.

Tones, shown in scheme (B), are refined, muted, and subtle and relate more to the winter season. They have a natural affinity for gray, because they contain it. Any and all tones can be put together effectively, although recommended combinations will be found on the lower schemes of Plate V and Plate VII.

Shades, shown in scheme (C), are dark, rich, and dignified. They have a natural affinity for black, because they contain it. Shades are associated with the browns, maroons, and purples of autumn. While any and all shades blend well together, suggested effects are visualized in the lower schemes of Plate IV and Plate VI.

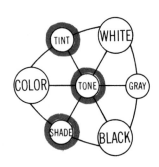

PLATE I

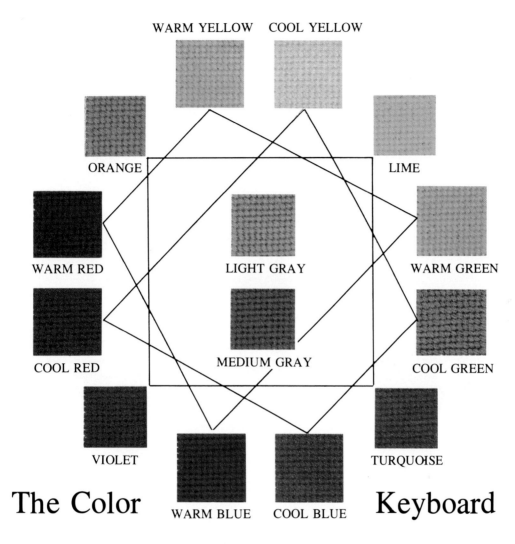

WARM YELLOW   COOL YELLOW

ORANGE   LIME

WARM RED   LIGHT GRAY   WARM GREEN

COOL RED   MEDIUM GRAY   COOL GREEN

VIOLET   TURQUOISE

The Color   Keyboard

WARM BLUE   COOL BLUE

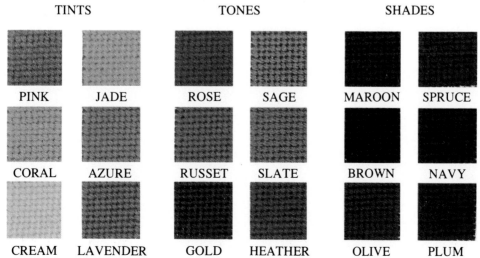

TINTS		TONES		SHADES	
PINK	JADE	ROSE	SAGE	MAROON	SPRUCE
CORAL	AZURE	RUSSET	SLATE	BROWN	NAVY
CREAM	LAVENDER	GOLD	HEATHER	OLIVE	PLUM

PLATE II

Palettes of Distinction

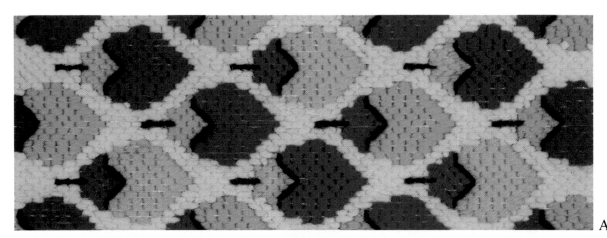

A

Warm red, warm yellow, warm green, warm blue, black, white

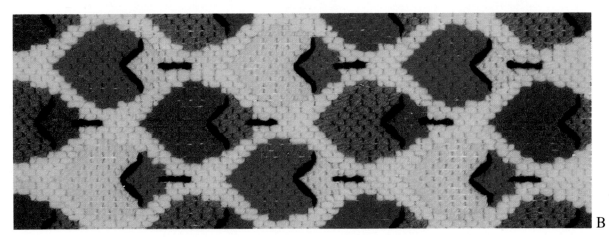

B

Cool red, cool yellow, cool green, cool blue, white, black

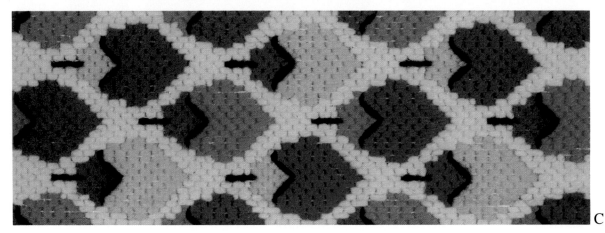

C

Orange, lime, turquoise, violet, white, black

PLATE III

Tints, Tones, and Shades

Pink, coral, cream, jade, azure, lavender, white *delicate - flowerlike*

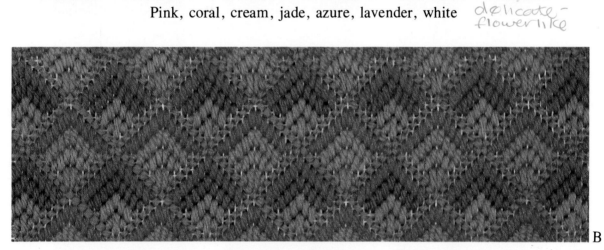

Rose, russet, gold, sage, slate, heather, medium gray *refined - subtle*

Maroon, brown, olive, spruce, navy, plum, black

PLATE IV

Analogous Colors: Brights and Shades

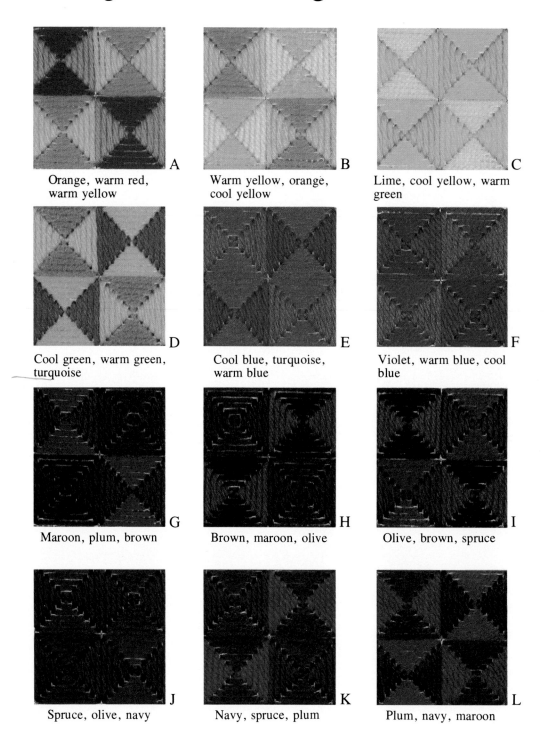

A — Orange, warm red, warm yellow

B — Warm yellow, orange, cool yellow

C — Lime, cool yellow, warm green

D — Cool green, warm green, turquoise

E — Cool blue, turquoise, warm blue

F — Violet, warm blue, cool blue

G — Maroon, plum, brown

H — Brown, maroon, olive

I — Olive, brown, spruce

J — Spruce, olive, navy

K — Navy, spruce, plum

L — Plum, navy, maroon

PLATE V

Analogous Colors: Tints and Tones

Key color appears twice *(handwritten note)*

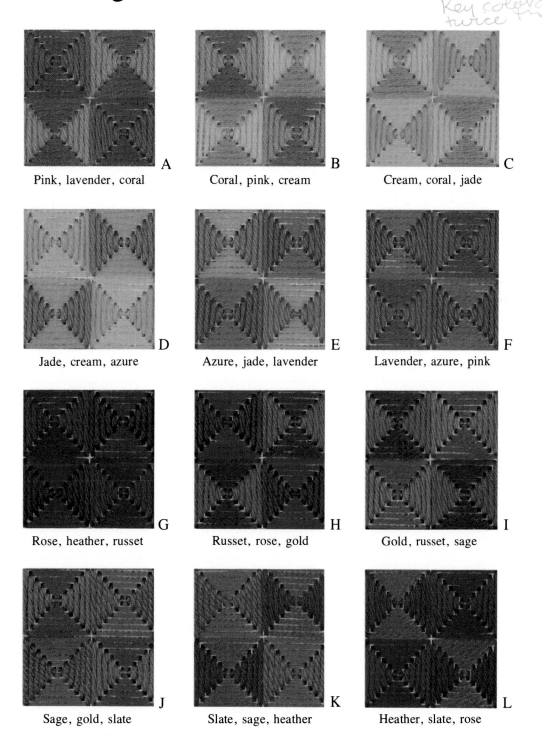

A — Pink, lavender, coral

B — Coral, pink, cream

C — Cream, coral, jade

D — Jade, cream, azure

E — Azure, jade, lavender

F — Lavender, azure, pink

G — Rose, heather, russet

H — Russet, rose, gold

I — Gold, russet, sage

J — Sage, gold, slate

K — Slate, sage, heather

L — Heather, slate, rose

PLATE VI

Complements: Brights and Shades

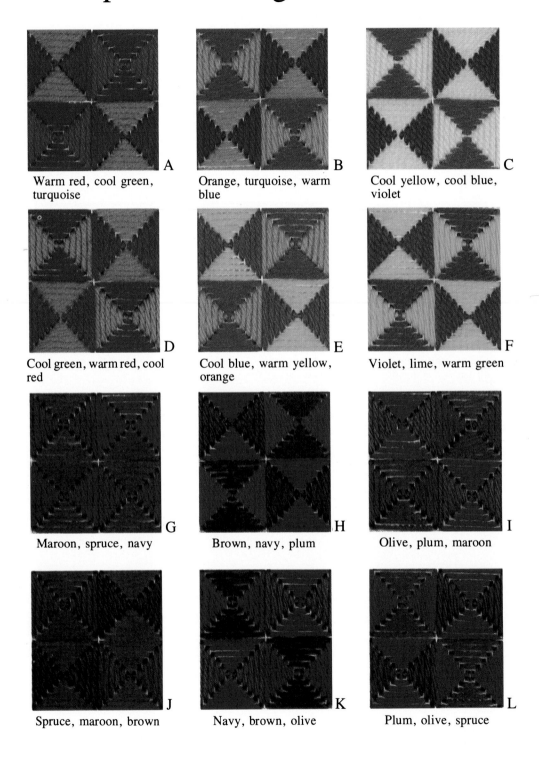

A
Warm red, cool green, turquoise

B
Orange, turquoise, warm blue

C
Cool yellow, cool blue, violet

D
Cool green, warm red, cool red

E
Cool blue, warm yellow, orange

F
Violet, lime, warm green

G
Maroon, spruce, navy

H
Brown, navy, plum

I
Olive, plum, maroon

J
Spruce, maroon, brown

K
Navy, brown, olive

L
Plum, olive, spruce

PLATE VII

Complements: Tints and Tones

sweet youthful

impression painting

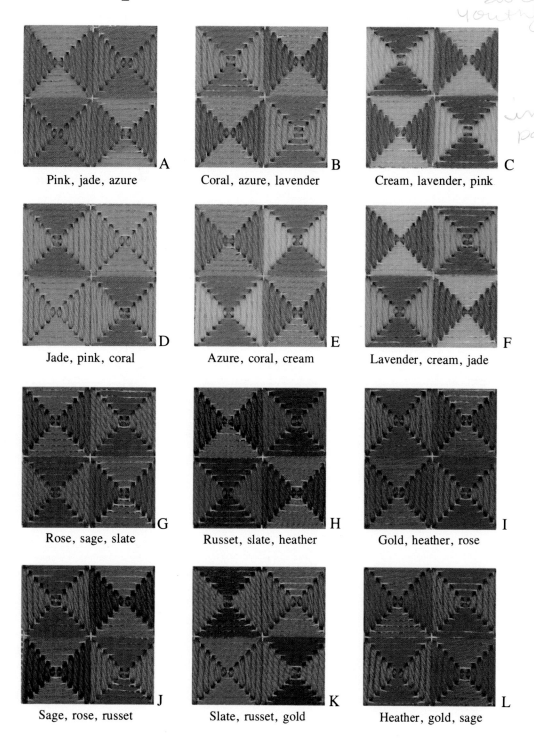

A Pink, jade, azure	**B** Coral, azure, lavender	**C** Cream, lavender, pink
D Jade, pink, coral	**E** Azure, coral, cream	**F** Lavender, cream, jade
G Rose, sage, slate	**H** Russet, slate, heather	**I** Gold, heather, rose
J Sage, rose, russet	**K** Slate, russet, gold	**L** Heather, gold, sage

PLATE VIII

Color to White: Color to Black

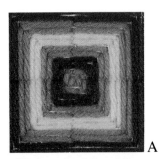

A

Warm red, pink, white

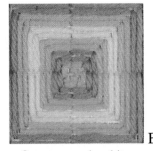

B

Orange, coral, white

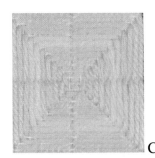

C

Cool yellow, cream, white

impressionistic painting sweet youthful

D

Cool green, jade, white

E

Cool blue, azure, white

F

Violet, lavender, white

Old Masters somber

G

Warm red, maroon, black

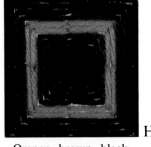

H

Orange, brown, black

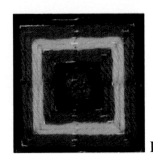

I

Lime, olive, black

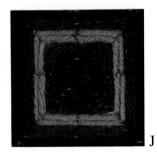

J

Cool green, spruce, black

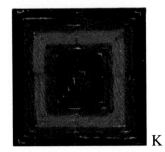

K

Cool blue, navy, black

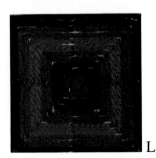

L

Violet, plum, black

PLATE IX

Tint to Shade: Color to Gray

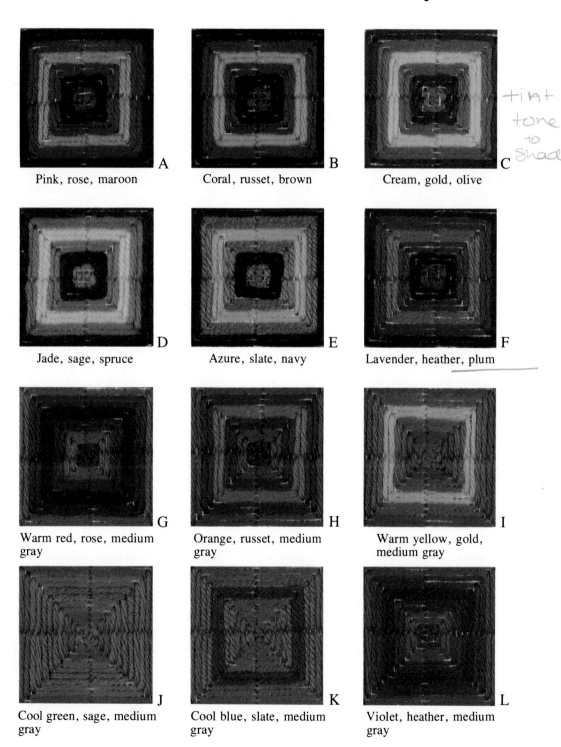

A — Pink, rose, maroon

B — Coral, russet, brown

C — Cream, gold, olive

tint tone to shade

D — Jade, sage, spruce

E — Azure, slate, navy

F — Lavender, heather, plum

G — Warm red, rose, medium gray

H — Orange, russet, medium gray

I — Warm yellow, gold, medium gray

J — Cool green, sage, medium gray

K — Cool blue, slate, medium gray

L — Violet, heather, medium gray

PLATE X

The Bright Grounds

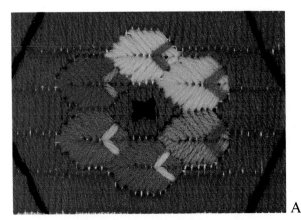

A

Warm red, warm yellow, warm green, cool green, turquoise, cool blue, violet, black

B

Orange, cool yellow, cool red, violet, cool blue, turquoise, cool green, black

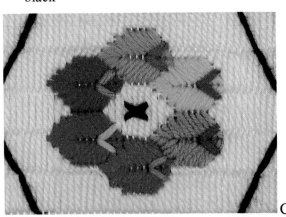

C

Cool yellow, orange, warm green, cool green, turquoise, violet, warm red, black

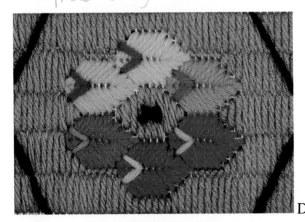

D

Cool green, warm yellow, orange, warm red, warm blue, turquoise, warm green, black

E

Turquoise, warm yellow, warm green, cool blue, violet, warm red, orange, black

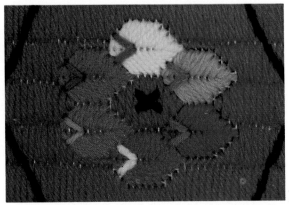

F

Warm blue, cool yellow, orange, warm red, violet, cool blue, turquoise, black

PLATE XI

The Tint Grounds

A

Pink, cream, rose, lavender, slate, jade, gold, white *Sweetness femininity*

B

Coral, cream, russet, pink, rose, azure, jade, white *fleshy - intimate (coral)*

C

yellow Cream, lime, warm green, sage, lavender, coral, gold, white *Spring - cheerful*

D

green Jade, cream, coral, rose, azure, lavender, jade, white *cool friendly*

E

Azure, pale azure, lavender, pink, coral, cream, jade, white *good for background for warm colors - most popular*

F

Lavender, cream, rose, pink, heather, coral, russet, white *refined royal exclusive*

27

PLATE XII

The Tone Grounds

softness gray nowhite

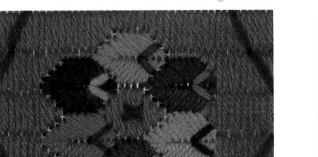

A

Rose, gold, olive, slate, navy, heather, brown, maroon

B

Russet, gold, maroon, heather, navy, slate, olive, brown

C

Gold, russet, sage, slate, cool blue, heather, rose, olive

D

Sage, gold, brown, rose, plum, slate, spruce

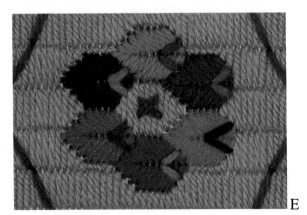

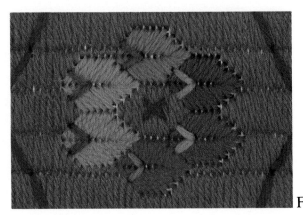

E

Slate, gold, spruce, cool blue, plum, rose, brown, navy

F

Heather, gold, russet, rose, violet, slate, sage, plum

28

PLATE XIII

The Shade Grounds

autumn — luminous

A

Maroon, gold, olive, cool green, navy,
turquoise, plum, rose

B

Brown, gold, cool green, navy, violet,
maroon, russet·

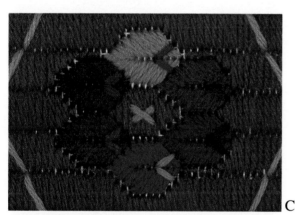

C

Olive, gold, spruce, navy, plum,
maroon, brown

D

Spruce, warm green, navy, turquoise,
warm blue, cool green, plum, sage

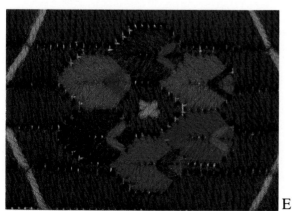

E

Navy, spruce, turquoise, warm blue,
violet, maroon, cool blue, slate

moonlight

F

Plum, maroon, warm green, warm blue,
cool green, heather, warm red

29

PLATE XIV

The Entire Keyboard

PLATE XV

unifying color

General Color Blending

A

Lime, olive, orange, brown, warm red,
maroon, cool blue, navy, white

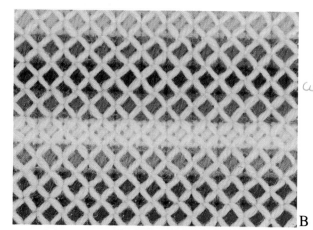

white

B

Coral, russet, brown, heather, cream,
gold, olive, violet, white *snowy*

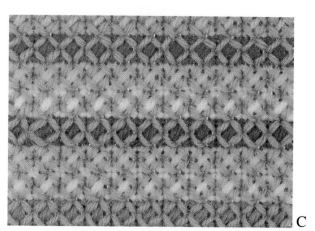

C

Pink, rose, lime, coral, russet, cream,
gold, light gray *Iridescent*

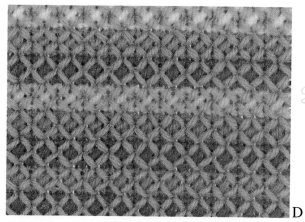

gray

D

Jade, sage, rose, azure, slate, lavender,
heather, light gray *gray fog*

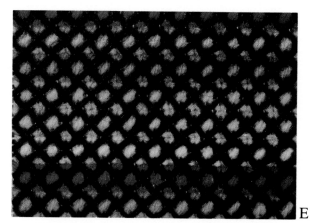

E

Warm red, pink, cool green, jade, warm
yellow, cream, warm blue, azure, black

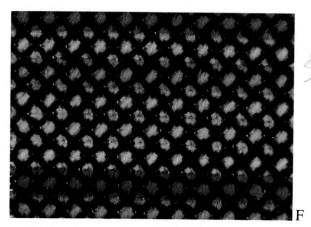

black sparkle

F

Turquoise, azure, orange, coral, warm
green, lime, violet, lavender, black

PLATE XVI

Optical Color Mixtures

A — Warm yellow, warm blue

not right —
need closer value or brightness

B — Warm red, cool blue

C — Maroon, plum

vibrates

D — Cool green, turquoise

E — Orange, warm green

F — Orange, russet

G — Gold, russet

H — Gold, olive

I — Brown, spruce

iridescent
now

J — Coral, light gray

K — Pink, lavender

L — Jade, light gray

Plate IV.
Analogous Colors:
Brights and Shades

All the colors shown in Plate IV are drawn from the color keyboard shown in Plate I. (This is true of all the color schemes shown in this book.) Many theorists on color and color harmony have concluded that colors are most attractive either (a) when they are closely related in hue or (b) when they are contrary and complementary. The great M. E. Chevreul agreed with this, and so did such later theorists as Munsell and Ostwald.

In considering the color schemes on page 20—and throughout the color plates—the reader would do well to keep the color circle and the color triangle in mind. This writer has used them as guides and as organized references rather than merely selecting colors at random. Once good color order is accepted as a starting point, the textile artist may move in any desired direction.

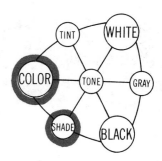

The six color schemes at the top of Plate IV (A, B, C, D, E, and F) consist of analogous bright, pure colors from the color keyboard of Plate I: orange with warm red and warm yellow (A); warm yellow with orange and cool yellow (B); lime with cool yellow and warm green (C); cool green with warm green and turquoise (D); cool blue with turquoise and warm blue (E); and violet with warm blue and cool blue (F). The key color in each case has been given twice the area of its adjacents. Other such analogous schemes can be plotted.

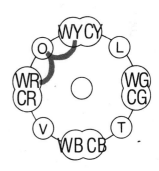

The six color schemes at the bottom of Plate IV (G, H, I, J, K, and L) are similarly arranged with analogous colors, but here all are shades: maroon with plum and brown (G); brown with maroon and olive (H); olive with brown and spruce (I); spruce with olive and navy (J); navy with spruce and plum (K); and plum with navy and maroon (L). Again, the key colors have twice the area of the adjacents.

Plate V.
Analogous Colors:
Tints and Tones

The reader will note that, when a color scheme is held to one specific form (bright hue, tint, shade, or tone), as in Plates IV, V, VI, and VII, it naturally looks harmonious. This is because the colors have visual elements in common. Pure, bright colors àre impulsive and dynamic. Tints are delicate. Shades are heavy and strong. Tones are restrained and refined. Each *form* thus has a unique emotional quality that is independent of *hue* itself.

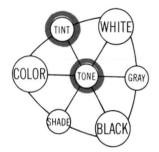

Analogous tints are presented at the top of Plate V, shown on page 21: pink with lavender and coral (A); coral with pink and cream (B); cream with coral and jade (C); jade with cream and azure (D); azure with jade and lavender (E); and lavender with azure and pink (F). (A softer pink might also be desirable.) The key color in each scheme has been given twice the area of its adjacents.

At the bottom of Plate V are shown analogous combinations of soft tones: rose with heather and russet (G); russet with rose and gold (H); gold with russet and sage (I); sage with gold and slate (J); slate with sage and heather (K); and heather with slate and rose (L). Note the double area for the key colors. The subtle beauty of these analogous combinations of tones is quite evident and lends itself to effects of rare dignity and elegance.

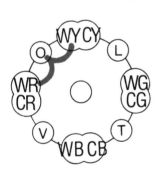

There is an important fact to note about analogous arrangements of color. They always favor one limited region of the spectrum or color circle: they are either dominantly warm or dominantly cool. This endows them with an *emotional* quality, for the simple reason that each adjacent contains elements of the key hue. (With complements the contrary is true, as is explained in connection with Plates VI and VIII.) It should again be emphasized that colors and color arrangements can have precise personalities.

Plate VI.
Complements:
Brights and Shades

If analogous colors have an emotional quality, complementary colors have an exciting *visual* quality, for they put contrary hues into direct conflict. Direct complements (pure hues) lie directly across from each other on the color keyboard shown in Plate I: warm red and cool green; orange and turquoise; warm yellow and cool blue; cool yellow and warm blue; lime and violet; and warm green and cool red. If a two-color scheme is desired, any of these pairs may be considered.

For contrast effects split complements are suggested. In this case a key hue is combined either with its direct opposite and an adjacent of this opposite or with the two colors that lie on either side of its direct complement. Slight variations from this principle are allowed, as long as the colors are visually startling in combination.

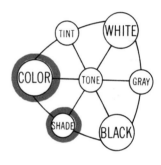

The upper schemes of Plate VI, shown on page 22, are composed of bright, pure hues: warm red with cool green and turquoise (A); orange with turquoise and warm blue (B); cool yellow with cool blue and violet (C); cool green with warm red and cool red (D); cool blue with warm yellow and orange (E); and violet with lime and warm green (F). The key hue in each scheme has been given twice the area.

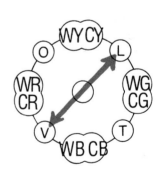

The lower schemes of Plate VI are composed of shades: maroon with spruce and navy (G); brown with navy and plum (H); olive with plum and maroon (I); spruce with maroon and brown (J); navy with brown and olive (K); and plum with olive and spruce (L). No one will deny the somber beauty of these combinations of shades. They are reminiscent of many old-world textiles, wool tweeds, plaids, and shawls of both western and eastern traditions. Though they involve complements, their depth and richness bespeak a cultured taste in color selection.

Plate VII.
Complements:
Tints and Tones

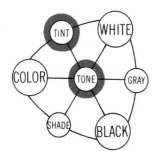

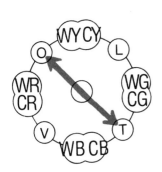

To compare analogous schemes and complementary schemes, it may be of interest to study Plates IV and VI for bright hues and shades and Plates V and VII for tints and tones. Although similar or identical colors are involved, their arrangements are different and thus their appearances are distinctly different.

Complementary tints lie at the top of Plate VII, shown on page 23: pink with jade and azure (A); coral with azure and lavender (B); cream with lavender and pink (C); jade with pink and coral (D); azure with coral and cream (E); and lavender with cream and jade (F). Rather sparkling effects are created, some of which have a ''sweet'' and youthful look.

Complementary tones are included at the bottom of Plate VII: rose with sage and slate (G); russet with slate and heather (H); gold with heather and rose (I); sage with rose and russet (J); slate with russet and gold (K); and heather with gold and sage (L). All six tones from Plate I are represented but in different arrangements. In all cases the key color is given double area.

Further harmonies and sequences of color—all from Plate I—will be found in the following plates. To review color schemes at this point, color effects have been derived chiefly from the color keyboard and the tints, tones, and shades shown in Plate I. Schemes that combine these basic forms have not yet been exhibited. Although the number of colors used is thirty, plus black, white, and one gray, some fifty-four arrangements have been shown.

If the colors included in Plate I are looked upon as a comprehensive color ''scale'' with sharps and flats, one can readily understand that ''songs'' and ''pieces'' of almost endless variety can be composed to satisfy virtually all human tastes.

Plate VIII.
Color to White:
Color to Black

The color schemes shown in Plate VIII and Plate IX are based on the color triangle previously described in this book. As already noted, straight paths on this triangle automatically lead to beauty, because the pure hues, white, black, tint, tone, shade, and gray are arranged in accordance with natural visual and psychological order.

The six schemes shown at the top of page 24 consist of a pure hue, a tint, and white, forms that have elements in common: warm red, pink, and white (A); orange, coral, and white (B); cool yellow, cream, and white (C); cool green, jade, and white (D); cool blue, azure, and white (E); and violet, lavender, and white (F). Much of Impressionistic painting in France followed this organization, using pure pigments mixed with white on a white canvas. This observation will be made again in the pages following the color plates.

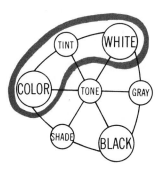

The six upper schemes are monochromatic, each based on a single hue. Other arrangements could be planned: analogous arrangements, such as warm red, coral, and white; cool green, cream, and white; violet, pink, and white; and so on around the color circle, or complementary arrangements, such as cool red, jade, and white; turquoise, coral, and white; or violet, lime, and white.

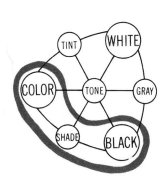

The six schemes at the bottom of Plate VIII consist of a pure hue, a shade, and black, also forms with common elements: warm red, maroon, and black (G); orange, brown, and black (H); lime, olive, and black (I); cool green, spruce, and black (J); cool blue, navy, and black (K); and violet, plum, and black (L). Variations of the sequence from pure hue to shade to black could be developed for analogous colors or complementary colors. Many of the Old Masters, such as Rembrandt, favored this somber path on the triangle, as will be noted later.

Plate IX.
Tint to Shade:
Color to Gray

The handsome schemes shown on page 25 are quite special. They also trace straight paths on the color triangle. At the top are sequences from tint to tone to shade: pink, rose, and maroon (A); coral, russet, and brown (B); cream, gold, and olive (C); jade, sage, and spruce (D); azure, slate, and navy (E); and lavender, heather, and plum (F). Each of these schemes is monochromatic and contains the same hue in different forms. Analogous colors could also be planned, such as cream, sage, and navy; coral, rose, and plum; or any other such combination of adjacents. Or the sequence could be in terms of split complements, such as cream, slate, and plum; coral, spruce, and navy; and the like.

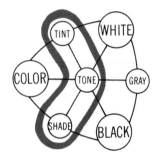

The tint-tone-shade combination offers subtle beauty. It is found, in principle, in the chiaroscuro of the Renaissance and was greatly admired by Wilhelm Ostwald, who called this scale the shadow series.

The bottom six schemes trace a path from pure hue to tone to shade, also a unique sequence: warm red, rose, and medium gray (G); orange, russet, and medium gray (H); warm yellow, gold, and medium gray (I); cool green, sage, and medium gray (J); cool blue, slate, and medium gray (K); and violet, heather, and medium gray (L). The arrangements may also be based on adjacents, such as warm red, heather, and medium gray; cool green, slate, and medium gray; and so on. Or opposites could be combined, such as orange, slate, and medium gray; cool yellow, heather, and medium gray; and so on.

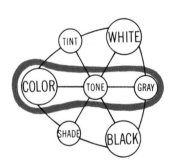

The great English painter Turner achieved remarkably luminous effects by blending strong tints or pure colors and shading them into neutral gray. In summary, sequences from pure color to black tend to appear lustrous. Shadings from pure color to gray tend to appear iridescent.

Plate X.
The Bright Grounds

Six pure hues from the color keyboard shown in Plate I and all six tints, tones, and shades are shown in harmonious color combinations in Plates X, XI, XII, and XIII. In most schemes six or seven colors are used and are listed with each combination. The colors of Plate X, shown on page 26, are limited essentially to pure hues and black.

On the four plates that involve colored backgrounds the personalities of individual pure hues, tints, tones, and shades are given a chance to predominate. Each scheme therefore has its own individuality. This writer ventures to guess that, especially with Plate X, strong likes and dislikes will be expressed by the reader. As a digression of considerable human interest, following are brief notes that review the emotional and psychological significance of personal color preferences.

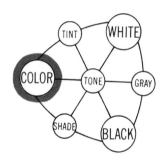

A preference for red (A) usually indicates a person with strong worldly interests, one whose attentions are directed outward into the world. Such a person may experience many shifts in mood.

Orange personalities (B) often have a social bent and are friendly, likable, and able to get along with others in all walks of life. Orange is a warm, luminous, and convivial hue.

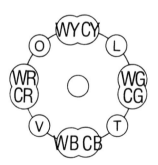

Yellow (C) is frequently associated with intellectuals, with persons attracted to oriental philosophies who delight in innovations and anything new.

Green (D) is a conventional color that indicates persons who take an active part in community affairs, enjoy parties, join clubs, and lead exemplary home lives.

Persons who like turquoise, or blue-green (E), are quite likely to have good taste and charming manners but to be vain and aloof from ordinary mortals.

Blue (F) is the color of self-control, introspection, inwardly directed interests. It is the token of hard work, application, success, and prosperity.

For further observations on color and people consult the author's *Color in Your World*.

Plate XI.
The Tint Grounds

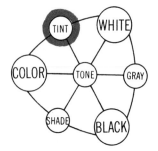

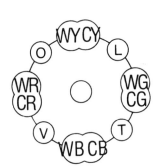

All the colors in the six schemes of Plate XI, shown on page 27, are limited to tints, some tones, and white. These are listed under each combination. Incidentally, the pink, coral, cream, and jade tints are slightly paler than those shown in Plate I.

In music the notes of the diatonic scale are few in number. Their vibrations are repeated in mathematical progression from octave to octave. Yet, despite this limitation, music of infinite variety can be written.

Terms for music and color are freely associated—chromatic, tone, scale, and harmonic, for example. Many musicians have related color to musical notes, but no scientific correlation of frequency of vibration between the two has ever been clearly established. Color, too, has an octave in the spectrum, in which relatively few *pure* hues can be distinguished. However, if white and black are added to pure colors (to make tints, tones, and shades), the number of discernible variations is vastly increased. Thus, color harmonies may have even greater potential than musical harmonies.

Each of the tint backgrounds shown in Plate XI has a beauty of its own. Pink (A) exhibits ''sweetness'' and femininity. Coral (B) has a ''fleshy'' quality and is one of the most intimate of all colors. Cream, or pale yellow (C), is like a spring daffodil, light, sunny, and often considered the most cheerful of colors. Jade, or pale green (D), is cool and fresh, natural, livable and friendly.

Pastel blue (E) is one of the most popular of all tints and is universal in its appeal. It tends to recede like the sky, an effective background for the display of warmer colors.

Lavender (F) is refined, royal, and exclusive. Its subtlety perhaps lies in the fact that it blends the two ends of the spectrum—red and blue—and, with white added, becomes a fairly exotic color that appeals to persons of discriminating and mature taste, such as artists and the culturally minded.

Plate XII.
The Tone Grounds

The six schemes shown on page 28 are limited to tones and shades. The colors used are listed under each combination. The common element here is softness and grayness, with no white or pure hue whatsoever.

To continue with analogies between color and music, loud music is associated with brilliant hue; light notes with tints; deep notes with shades; and soft notes with tones—relationships with which most persons will agree. Indeed, composers such as Wagner and Scriabin worked in terms of color. Arthur Bliss wrote a color symphony, and in *Fantasia* Walt Disney conveyed musical notes and rhythms in colors and abstract, animated designs.

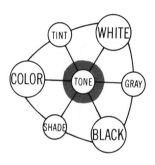

Many commercial textiles, needlecraft kits, and weaving patterns all too often display color schemes conventionally and academically. There is much "dipping" into the spectrum, with a little of this and a little of that. There may be a confusion of pure hues with pastel tints, offset with some tones and shades. Most persons are so familiar with these that the lack of good aesthetic harmony isn't noticed. The great majority of color effects look alike, lack originality, and confess to an amateurish quality.

Color undoubtedly has intrinsic appeal. So does the song of a bird, the patter of rain, and the babbling of an infant. Perhaps art is involved here, but since the beginning of history people have sought to create beauty in *human* rather than natural terms.

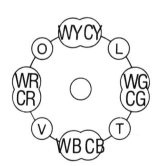

The color schemes of Plate XII, carried out with tone grounds and tone or shade accents, are founded on human concepts of beauty. Some may be reminiscent of nature, but nature has by no means been copied. Each of the grounds—rose (A), russet (B), gold (C), sage (D), slate (E), and heather (F)—all drawn from Plate I, has a character of its own. They exhibit distinction and a precise overall harmony and not merely a number of colors naively put together.

Plate XIII.
The Shade Grounds

Shades with some pure colors are used for the six schemes shown on page 29. All the grounds are shades from Plate I. The reader will note that these effects have a deeper richness than schemes with tone grounds, as on Plate XII. There is one further and impressive difference that will hopefully be obvious without too much explanation.

There are in the world color effects that depend on the quality of illumination. Daylight, for example, may have a pink and orange cast at dawn, scaling to yellow, white, and the clear blue of a midsummer day. There are also the mellow glow of Indian summer, the weird grayish and greenish light that sometimes precedes a thunderstorm, the purple mist of great distance, and the blueness of night. The writer feels that some of these extremely harmonious effects have been caught in the combinations of Plate XIII.

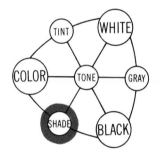

Grounds in maroon (A), brown (B), and olive (C) have the "feel" of autumn. Each suggests a predominantly warm illumination in rose, orange, and yellow-green. To a fairly convincing extent these schemes might be looked upon as being composed of *normal* colors seen under a colored light source. Does the reader agree?

With a spruce ground (D) the "colored-light" effect is quite apparent. The leaf forms of the pattern might well be perceived as *normal* colors seen under the influence of green light. All the colors blend harmoniously together, within a keyed green atmosphere.

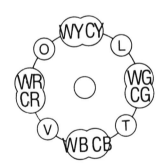

A navy ground (E) has the appearance of night or moonlight. All the colors are dominantly influenced by blue, and there is no inconsistent or jarring note.

A plum ground (F) might convey the aspect of purple mist within which lime, green, and red lights shine forth.

If these luminous impressions are acceptable to the reader, the art of color is called upon to create new expressions.

Plate XIV.
The Entire Keyboard

If the textile artist is able to plan color schemes with a well-organized array of colors—such as those in the color keyboard of Plate I—the problem of harmony is greatly simplified. There are visual (and emotional) relationships among pure hues and their corresponding pastel tints, rich shades, and muted tones.

The designer of Plate XIV, shown on page 30, was given full reign in terms of pattern and color combinations. The only restrictions were (a) to adhere to the colors shown in Plate I and (b) to use all thirty of them, plus two grays, a black, and a white. The results are as shown.

Perhaps further analogies may be drawn between color and music. The several "patches" in Plate XIV may be compared to different keys, for example, or to different tempos, such as fortissimo for strong accents and high contrast, pianissimo for a group of delicate pastels, allegro for visually brisk effects such as split complements, and adagio for combinations of deep shades and low contrast. Perhaps the reader can see other qualities, musical or not.

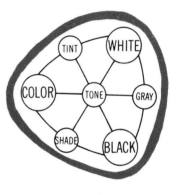

A great number of textile color harmonies tend to look alike. Color, of course, has intrinsic appeal in and of itself. A small child, given a box of crayons, will glory in them, eagerly trying one after another until the box is exhausted. The warmer hues are normally the first choices, with brown, gray, purple, and black following.

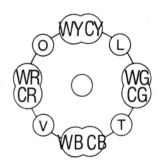

Yet as training, education, and culture are absorbed, more discrimination is shown and, most unusual of all, personality traits develop. People become outgoing, conservative, bold, demure, social, reclusive—and show countless other shades of human temperament. Each may have different ideas about color harmony, and each should, for self-satisfaction, arrange colors that are temperamentally compatible.

Plate XV.
General Color Blending

Any group of colors, gaudy, discordant, or otherwise, can often be brought into attractive harmony if there is a pervading unifying element to "pull everything together." This is exemplified in the schemes shown on page 31, which call upon white, gray, and black as unifying forces. Such harmony may be accomplished through cross-stitch, embroidery, crewel, and weaving techniques, as demonstrated.

In scheme (A) contrasting colors, lime and olive, orange and brown, warm red and maroon, and cool blue and navy, are harmonized with a white "net" spread over them. In scheme (B) the white "net" harmonizes a rather odd combination of coral, russet, brown, heather, cream, gold, olive, and violet, which in themselves could hardly be considered well coordinated. The general result has a "snowy" or "crystallike" quality.

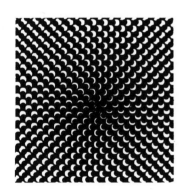

In schemes (C) and (D) the unifying color is light gray. In (C) a miscellaneous combination of pink, rose, lime, coral, russet, cream, and gold is softened to a mellow charm that may appear somewhat iridescent. Scheme (E) contains more restrained colors, jade, sage, rose, azure, slate, lavender, and heather. The harmony resembles the beauty of colors seen through mist on a cloudy day.

While white and gray were used for unity in schemes (A), (B), (C), and (D), light colors can also be considered, and many more effects can be devised. Recommended as "nets" would be such tints as cream, coral, and jade or even intense hues such as warm red, orange, cool yellow, and warm green.

In schemes (E) and (F) the "net" is black. In (E) primary hues are featured: warm red, pink, cool green, jade, warm yellow, cream, warm blue, and azure. (F) consists of secondaries: turquoise, azure, orange, coral, warm green, lime, violet, and lavender. Because of excessive differences in brightness and the strong influence of black, a fairly lustrous result is achieved—the pattern colors tend to sparkle.

Plate XVI.
Optical Color Mixtures

Most women are familiar with the shimmering beauty of changeable silks. These textiles, seen in folds, have a different glint in highlight, in shadow, and when viewed from different angles. The principle behind this involves optics and can be directly explained. Optical mixtures of color formed the basis of the "divisionism" techniques of Neo-Impressionist painters such as Seurat and Signac, who around the turn of the century applied their pigments in small dots in order to force them to be blended on the retina of the eye.

The French writer Chevreul had already dealt with the phenomenon. Optical mixtures were most effective when the colors involved *had fairly uniform brightness or value*. If values were excessive, results were poor. Chevreul himself was a textile colorist, and his experiments with *textiles* influenced the art of French *painting*.

Scheme (A) on page 32, which consists of cream and warm blue, represents what should *not* be done: the two colors are too different in brightness and the result does not have much effect. In scheme (B), which consists of warm red and cool blue, and scheme (C), based on maroon and plum, the optical blend lives and vibrates.

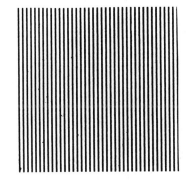

Other suggestions are shown: cool green with turquoise (D); orange with warm green (E), which "mixes" to a golden yellow; orange with russet (F); gold with russet (G); gold with olive (H); and brown with spruce (I), which is particularly rich.

Across the bottom of Plate XVI are other schemes: coral with light gray (J); pink with lavender (K); and jade with light gray (L). In these the blend of tinted color against a neutral gray of similar value results in iridescence.

Optical color mixtures, done in any technique and with any kind of yarn, may be considered for "yard goods" by themselves or as backgrounds on which colorful patterns can be superimposed. They bring the art of color to greater heights, introducing refinements that demand skill but increase the talents of those who seek to become competent textile artists.

Formal Color Harmony

This latter part of the book concerns formal principles of color harmony as conceived by more technically minded theorists. Traditional color harmony, as often taught in schools and colleges, is reviewed. Numerous books have been devoted to color harmony, and important ones are listed and described in the Suggested Reading section at the end of the book.

As to textiles themselves, a history can be found in many excellent volumes, and a few recommended ones are also included in the Suggested Reading list. This writer has decided to follow a discussion of formal color harmony with a chapter on the artist's viewpoint and to feature the writings of great painters since the Renaissance on color and on how they have used color. Hopefully, such observations will be new to the reader or at least encouraging to artists working with yarns and threads who seek to achieve outstanding color effects. Wherever possible the unique color arrangements of great painters have been demonstrated on some of the plates; the reader will be advised in such cases.

The sixteen color plates exhibit examples of harmony drawn from both a special color circle (Plate I) and the color triangle previously described. Let me now discuss *formal* principles of color harmony and follow them with remarks on the intuitive views of artists.

Probably the first and surely the most famous conclusions were drawn by M. E. Chevreul in France. In 1839 he published one of the greatest books on color, *The Principles of Harmony and Contrast of Colors*. Chevreul was a member of French and British scientific academies. He supervised the production of dyestuffs at the renowned Gobelin tapestry works outside Paris. What he wrote about visual color phenomena strongly influenced

DA VINCI

CHEVREUL

French Impressionism and Neo-Impressionism, for which he was rewarded at the age of one hundred with a bronze medal struck in his honor, a statue, and a special reprint of his masterpiece by the French government.

It is through Chevreul that such concepts as the harmony of adjacents, opposites, split complements, triads, and tetrads entered the language of color. He further remarked that complementary colors were superior to all other arrangements. Colors that appear none too pleasing side by side could be improved by separating them with black or white outlines. A primary color (red, yellow, or blue) combined better with two secondaries (orange, green, or violet) than a secondary combined with two primaries. Black looks best with deep colors such as blue and violet, while white looks best with light colors such as yellow, orange, and green. If light and dark colors are used together, gray may be the ideal separation or ground. The color schemes shown in the plates have respected much of this advice.

Another well-known writer on the subject of color, who took a rather formal approach, was the German poet Goethe. His *Farbenlehre* (*Theory of Colors*) was published in 1810 in German and was followed by an English translation in 1840. Goethe argued with Newton's theories and strongly promoted a psychological and aesthetic attitude toward color.

Goethe had strong opinions about color. He noted the aversion of many painters of his time to color and hoped that "our labors may tend to diminish this prejudice." To him some colors were *plus:* yellow, orange, and red-orange. The *minus* colors were blue, violet, and purple. All colors had a decided personality.

Concerning harmony, Goethe thought yellow and blue were a poor combination. Yellow and red had "a serene and magnificent effect." Orange and purple were exciting and elevating. Curiously, he wrote, "The juxtaposi-

GOETHE

tion of yellow and green has always something ordinary, but in a cheerful sense; blue and green, on the other hand, is ordinary in a repulsive sense. Our good forefathers called the last fool's colors.''

The theory of Munsell is also widely taught in the United States. His system, however, is far more famous as a method of color notation and identification than as an analysis of beauty. In his day, the early part of the twentieth century, Munsell held conservative, late-Victorian prejudices. He suggested that ''Beginners should avoid strong color'' and that ''Quiet color is a mark of good taste,'' statements that hardly find modern approval.

Like Chevreul, Munsell liked complementary colors. But he had reservations: ''The sense of comfort is the outcome of balance, while marked unbalance immediately urges a corrective.'' His central gray scale had nine steps, and the middle step, 5, had particular magic. To him value (brightness) steps were best when they had a neat order—3, 5, 7 or 4, 5, 6, with step 5 the ideal point of balance. If complements were chosen, an intense color such as red should be used over a smaller area than a complementary blue-green. And if areas of the two colors, spun on a disk, equaled a neutral gray of value 5, all would be well.

MUNSELL

The problem with Munsell is that many of his original theories too often led to dull and somber results. Yet few systems have a better and more precise order.

While Munsell's color scales are based on vertical value steps and horizontal chroma (purity) steps, Ostwald organized color scales that related to qualities of whiteness, blackness, and grayness in color. He concluded that most assortments of color that have equal hue and equal white and black content (visual) would be harmonious, regardless of their values. Colors with equal black content but different white and hue content were harmonious; colors with equal white content but different black and hue content were also harmonious; most appealing of all,

OSTWALD

colors with an apparent equal hue content but with different black and white content were likened to the magnificent chiaroscuro harmonies of the Renaissance.

Ostwald's contributions are well exemplified in the scales of the color triangle, and harmonies derived from them are freely represented in the color plates of this book. There is a natural order of colors that scale parallel to pure color and white. The deeper values are pure, and the lighter values weak. There is a further natural order of colors that scale parallel to pure color and black. The lighter values are pure, and the deeper values weak. There is another natural order of colors that scale parallel to black and white. The colors have an apparently equal purity but differ in light and dark content.

There are further scales from tint to tone to black; from shade to tone to gray; and from pure hue to tone to gray. Examples of most of these scales are found in the color plates.

Let the reader take courage and accept these views as principles rather than laws. Color circles, color triangles, and color systems at best resemble the organized notes on a piano. With them it is possible to play anything from chopsticks to majestic concertos.

How talented are you?

The Artist's Viewpoint

The art of color as it is appreciated today had its beginning in the Renaissance of the fifteenth and sixteenth centuries. In older civilizations, Egypt, Asia Minor, the Orient, Greece, and Rome, the palettes used in painting and decoration had a simple order: red, gold, yellow, green, blue, purple, and pink. Few people realize that ancient color expression had much of its basis in symbolism. Practically all Egyptian art has the same color palette, as does Greek polychrome decoration. It is apparent that early artists followed the dictates of conventional symbolism and rarely departed from them.

Yet in the Renaissance two great innovations occurred. Oil painting made possible an unlimited array of color. Tapestries and brocades utilized new dyestuffs and evolved into textiles of striking magnificence. Color scaled heights of grandeur not achieved before.

Renaissance painting was extremely colorful, and much of it portrayed the lush textiles of the day. The painting of draped fabrics became a consuming interest and was often given more attention than the depiction of saints, royalty, and nobility. What were the chief colors used? Raphael, for example, took delight in the simple beauty of red, yellow (gold), and blue.

Many Renaissance painters had an exceptional sense of color. John Ruskin wrote of ''supreme colorists'' like Giorgione, Tintoretto, Correggio—and Titian. (Other Italians of merit would be Mantegna and Veronese, who portrayed *The Last Supper* extravagantly as a sumptuous feast replete with crystal, silver, jewels, silks, and satins.) Titian's paintings had a golden quality associated with the Venetian school. With gold as a dominant theme, Titian's coloring included orange, red, muted blue, olive green, brown, and a creamy white.

Leonardo da Vinci was one of the most ingenious and brilliant of all Italian painters. He invented the chiaroscuro style well represented in the top schemes of Plate IX. What is meant by chiaroscuro? Da Vinci mastered the plastic qualities of light and shade. To explain, pink is not a highlight of red. To mix color scales by adding white to pure color is untrue to nature and a practice commonly followed by the uninitiated artist. In chiaroscuro the apparent quality of hue is maintained from highlight to shadow. With red the highlight would be vermilion, not pink, and the shadows would be a soft maroon, not a rich crimson.

Da Vinci's technique revolutionized the art of painting and was followed for generations. Concerning color harmony he wrote: "Of different bodies equal in whiteness, and in distance from the eye, that which is surrounded by the greatest darkness will appear the whitest; and on the contrary, that shadow will appear the darkest which has

TITIAN

the brightest white around it. Of different colors equally perfect, that will appear most excellent which is seen near its direct contrary: a pale color against red; a black upon white . . . blue near a yellow; green near red: because each color is more distinctly seen, when opposed to its contrary, than to any other similar to it.'' This amazing statement has been repeated in essence by color theorists ever since!

In Toledo, Spain, the great painter El Greco had an unusual style and concept of color. His religious figures were elongated and seemed to be not of this earth. He used deep, rich colors (shades) and scaled them through tones into eerie and chalky whites. (Note this sequence on the color triangle.) His pale tints were weak in chroma (purity), and his deep shades strong in hue. Such a sequence is unnatural for the most part and probably for this reason gave El Greco's art a supernatural quality. As color notes he liked red, orange, green, blue, and yellow, blackish at one extreme and whitish at the other.

DA VINCI

Rembrandt's treatment of color was precisely the opposite of that of El Greco. His usual style was to have rich, pure colors shade into black or deep brown, the sequence on the color triangle being from pure hue to shade to black. Color effects of this sequence are seen in the bottom schemes of Plate VIII. With Rembrandt the colors of higher value are pure and the deep values blackish and brownish—with most of his canvases pervaded by a golden light. This luminous style was copied by other artists for generations.

The English painter Sir Joshua Reynolds was an admirer of the Italian Titian. In his day he wrote, "There is not a man on earth who has the least notion of coloring; we all of us have it equally to seek for and find out, as at present it is totally lost to the art." Reynolds deduced that beauty required a dominance of warm color in a composition: "It ought, in my opinion, to be indispensably observed that the masses of light in a picture be always of a warm mellow color, yellow, red, or a yellowish-white;

EL GRECO

53

and that the blue, the gray, or the green color be used only to support and set off these warm colors; and for this purpose a small portion of cool colors will be sufficient.'' It is said that Gainsborough deliberately reversed the principle and in so doing painted his masterpiece, *The Blue Boy*.

Great progress in the art of color came with the English J. M. W. Turner, one of the most brilliant colorists of all time. Turner achieved strikingly luminous effects, anticipated Impressionism, and significantly influenced color expression to this day. Although Turner lived at a time when realism was essential in art, he became so enthralled with color that he created compositions that were almost completely abstract. His original style was to tone pure colors and strong tints subtly toward medium gray—with an occasional black or dark accent. In effect he shifted his sequences on the color triangle from pure hue to tone to gray. Examples of this technique will be found in the schemes on the bottom of Plate IX.

Turner admired Goethe and based compositions on the German's theories. He also knew of the color circle of Moses Harris and did variations on it. If color arrangements are plotted from purity to neutrality, luminous or iridescent results can readily be perceived.

Preceding the original school of Impressionism in France were the influences of Turner, the masterwork of M. E. Chevreul on color contrast, and the inspired leadership of Eugène Delacroix. Delacroix was the hero of the Impressionists, a man of aristocratic bearing, a friend of the important personages of his day (Victor Hugo, Stendhal, Gautier, Dumas, Chopin, and George Sand), and a devout champion of color who declared, ''Give me mud, and I will make the skin of a Venus out of it, if you allow me to surround it as I please.'' He studied the principles of Chevreul and was worshipped by young painters such as Monet and Van Gogh; Manet asked permission to copy a Delacroix painting.

With Impressionism a new world of color was introduced that differed markedly from the past. Pure colors, now conveniently available in tubes, were mixed with white and applied boldly on white canvases. Earth colors were sometimes avoided. The sequence in the color triangle was pure hue, tint, and white, with an occasional deep accent. Examples of this style will be found in the upper schemes of Plate VIII.

With Neo-Impressionism science dominated the artists' expression. Pure colors, applied in small dots by painters such as Georges Seurat and Paul Signac, were meant to be mixed visually like tiny light sources. Earth colors were mostly avoided. Painters studied the work of scientists such as the German Hermann von Helmholtz, the English James Clerk Maxwell and the American Ogden N. Rood. Never before had art paid such homage to science.

TURNER

Seurat has intriguing ideas about color harmony and the relation of color to design or pattern. Gaiety of tone, for example, was expressed with warm colors (red) and with lines rising from the horizontal. Sadness of tone was encountered in cool colors (blue) and in lines descending from the horizontal. There was calmness and equality in a balance of light and dark, warmth and coolness, and horizontal lines.

The American Ogden N. Rood was a leading authority on physiological optics, visual color mixtures, color vision, and afterimages. His book *Modern Chromatics* (1879) became the bible of the Neo-Impressionists. He wrote of the harmony of opposites, adjacents, and triads and presented extended lists of bad, inferior, disagreeable, tolerable, and excellent color combinations. He introduced what later became fundamental to one of Munsell's theories of harmony, ''We return now to the proposition that the best effect is produced when the

DELACROIX

colors in a design are present in such proportions that a composite mixture of them would produce a neutral gray,'' a conclusion that is both limiting to creative expression and basically untrue.

Fauvism was the first art movement of the twentieth century. Associated with it were Henri Matisse, André Derain, and Maurice de Vlaminck. Color was blatantly and impulsively applied. The Fauvists used color like ''sticks of dynamite,'' as one critic put it. No formal principles were followed, but color as intense as could be imagined was the primary aim. Vlaminck declared, ''We were always intoxicated with color.''

Orphism followed in the work of Robert Delaunay, who after seventy years again went to Chevreul for counsel. With geometrics and abstract compositions Delaunay, ''Like a child with its favorite toy . . . took the rainbow to pieces and improvised with the separate parts.''

SEURAT

The idea of color as beautiful in itself, apart from realism and form, found noble expression in abstract, nonobjective art. The idea that color held intrinsic powers had an outstanding champion in Wassily Kandinsky. He not only was a great painter but also an exceptional teacher and writer: "Generally speaking, color directly influences the soul. Color is the keyboard, the eyes are the hammers, the soul is the piano with many strings. The artist is the hand that plays, touching one key after another purposively, to cause vibrations in the soul." To which he added in italics, *"It is evident that color harmony must rest ultimately on purposive playing upon the human soul; this is one of the principles of the internal necessity."*

This may be somewhat esoteric, but Kandinsky had a consuming interest in color: "Inner necessity is the basis of both small and great problems in painting. . . . The starting point is the study of color and its effects on men." In other words, color came first. He believed in disciplined and serious observation: "The artist must train not only his eye but also his soul, so that it can weigh colors in its own scale and thus become a determinant in artistic creation."

Like Goethe before him, Kandinsky had strong feelings about color. To him warmth and coolness were exemplified in yellow and blue. Yellow had a spreading action; blue moved in upon itself "like a snail retreating into its shell"; red was stable, powerful, and "rang inwardly."

The reader may enjoy comparing his own views of color with those of Kandinsky. To him yellow was an earthly color and never acquired much depth. Blue was a heavenly color and created a feeling of rest. Green, a mixture of yellow and blue, was a color that Kandinsky did not like. While green was restful, it was also boring and to him represented "the social middle class, self-satisfied, immovable, narrow." Orange "is like a man convinced of his own powers. . . . Violet . . . has a morbid extinct quality, like slag. . . . There remains brown, unemotional, disinclined to movement.'' White was like absolute silence; black was motionless like a corpse.

Not to end this book on a dour note, color is one of the most delightful pleasures of the senses. Reaction to it is emotional, psychic, and spiritual, and joy in it requires no thought, deliberation, or training. One merely has to be exposed to it, and response is spontaneous.

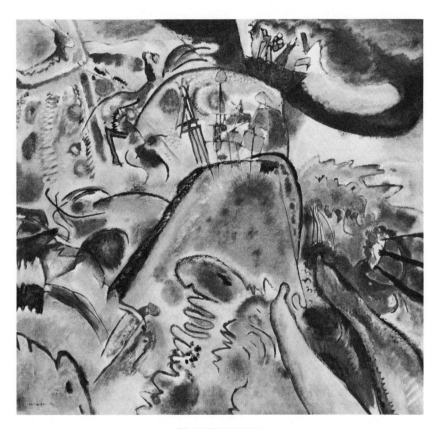

KANDINSKY

Acknowledgments

Page 50, Titian, *Pietro Arentino,* The Frick Collection, New York. Page 51, Da Vinci, *Virgin of the Rocks,* National Gallery, London. Page 52, El Greco, *The Adoration of the Shepherds,* Metropolitan Museum of Art, New York, Rogers Fund. Page 54, Turner, *Valley of Aosta—Snowstorm, Avalanche, and Thunderstorm,* Art Institute of Chicago, Frederick T. Haskell Collection. Page 55, Delacroix, *The Abduction of Rebecca,* Metropolitan Museum of Art, New York. Page 57, Seurat, *Entrance to the Harbor, Port-en-Bessin,* Museum of Modern Art, New York, Lillie P. Bliss Collection. Page 59, Kandinsky, *Little Pleasures, No. 174.* Solomon R. Guggenheim Museum, New York.

Suggested Reading

Birren, Faber, *Color and Human Response,* Van Nostrand Reinhold Company, New York, 1978. An original work devoted to biological, physiological, visual, emotional, and psychic reactions to color, with data on the use of color in healing and mental disturbance.

Birren, Faber, *Color, A Survey in Words and Pictures,* University Books, Secaucus, New Jersey, 1963. A general, historical, and modern review of interesting facts about color and its role in life through the ages.

Birren, Faber, *Color for Interiors,* Whitney Library of Design, New York, 1963. Color traditions in architecture and interior design, period styles, and the like. Contains charts of actual color chips.

Birren, Faber, *Color in Your World,* Collier Books, New York, 1978. An analysis of personality and character traits based on liked and disliked colors.

Birren, Faber, *Color Perception in Art,* Van Nostrand Reinhold Company, New York, 1976. Advanced principles of color expression in art described and illustrated in terms of modern inquiry into human perception.

Birren, Faber, *Color Psychology and Color Therapy,* University Books, Secaucus, New Jersey, 1977. A basic reference for over 25 years.

Birren, Faber, *Creative Color,* Van Nostrand Reinhold Company, New York, 1978. A basic course on color harmony used as a text in many art schools, colleges, and universities.

Birren, Faber, *History of Color in Painting,* Van Nostrand Reinhold Company, New York, 1980. Covers the subject thoroughly with over 300 pages and 500 illustrations.

Birren, Faber, *Light, Color, and Environment,* Van Nostrand Reinhold Company, New York, 1969. Reviews the role of color in architecture and period styles and suggests actual color schemes for homes, schools, hospitals, and offices.

Birren, Faber, *Principles of Color,* Van Nostrand Reinhold Company, New York, 1978. A fundamental book on color harmony, required reading for judges of the National Council of State Garden Clubs.

Birren, Faber, *The Story of Color,* Crimson Press, Westport, Connecticut, 1941. Long out of print but excellent reference on virtually all aspects of color: science, art, mysticism, healing, symbolism, mythology, and religion.

Chevreul, M.E., *The Principles of Harmony and Contrast of Colors* (1839), edited and annotated by Faber Birren, Reinhold Publishing Corporation, New York, 1967. One of the most famous books on color ever written, with a great influence on education and art to this day.

Gerritsen, Frans, *Theory and Practice of Color,* Van Nostrand Reinhold Company, New York, 1975. An excellent modern treatise on color, theoretical and practical.

Goethe, Johann Wolfgang von, *Theory of Colors,* translated by Charles Lock Eastlake (1840), M.I.T. Press, Cambridge, Massachusetts, 1970. Another key work in the literature of color, still fascinating to study.

Graves, Maitland, *The Art of Color and Design,* McGraw-Hill, New York, 1951. Devoted to the work of Albert Munsell, successfully used as a textbook in art and color education.

Itten, Johannes, *The Elements of Color,* foreword by Faber Birren, Van Nostrand Reinhold Company, New York, 1970. A Bauhaus textbook widely used for color training in art.

Jacobsen, Egbert, *Basic Color,* Paul Theobald and Company, Chicago, 1948. One of the best books on color, devoted to the remarkable work of Wilhelm Ostwald.

Jones, Tom Douglas, *The Art of Light and Color,* Van Nostrand Reinhold Company, New York, 1972. Concerned with lumia—luminous and kinetic color effects using light for beautiful mobile results.

Justema, William and Doris, *Weaving and Needlecraft Color Course,* Van Nostrand Reinhold Company, New York, 1971. One of the few books exclusively on this subject.

Kandinsky, Wassily, *The Art of Spiritual Harmony,* Houghton Mifflin Company, Boston, 1914. The great abstract-color artist's engaging ideas about the spiritual effects of color.

Matthaei, Rupprecht, *Goethe's Theory of Color,* translated by Herb Aach, Van Nostrand Reinhold Company, New York, 1970. A large and beautiful work on Goethe, heavily illustrated in color.

Munsell, A.H., *A Grammar of Color,* edited and annotated by Faber Birren, Van Nostrand Reinhold Company, New York, 1969. An excellent treatise on color harmony.

Ostwald, Wilhelm, *The Color Primer,* edited and annotated by Faber Birren, Van Nostrand Reinhold Company, New York, 1969. A translation and interpretation of Ostwald's original thoughts on color organization and harmony.

Rood, Ogden, *Modern Chromatics* (1879), edited and annotated by Faber Birren, Van Nostrand Reinhold Company, New York, 1973. By one of America's greatest authorities on color perception, an influence on Neo-Impressionism in France.

Index

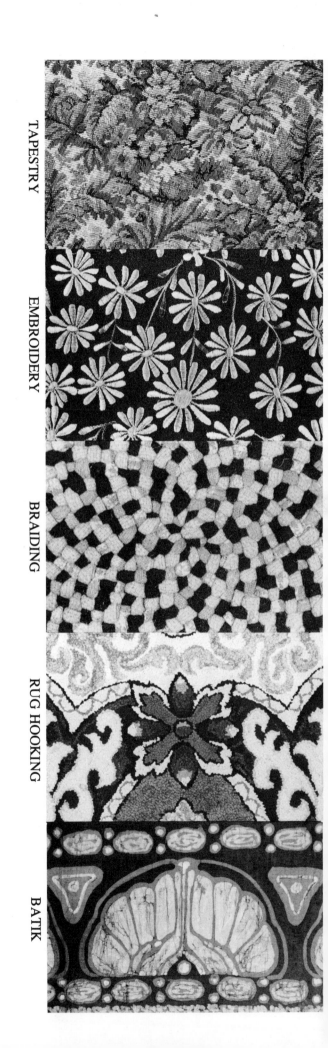

TAPESTRY

EMBROIDERY

BRAIDING

RUG HOOKING

BATIK